Biology at Princeton

(1947-2012)

John Tyler Bonner

Cover: The illustration of the life cycle of John Tyler Bonner was designed by the late Amy Bordvik as a poster in celebration of John's 90th birthday in 2010. Henry Horn drew the caricature of John Bonner's life cycle as that of his beloved research subject, a cellular slime mold. Photos of John as a larva and as a young scientist were scanned from the dust jacket of his autobiography *Lives of a Biologist* (2002, Harvard University Press, Cambridge Ma), and that of a later stage from his reflections on Life Cycles (1993, Princeton University Press, Princeton, NJ). John is on the left end of the middle row in the photo of the Princeton Biology Department Faculty 1948-9 (Orren Jack Turner Studio, Princeton, NJ). The "sultan of slime" photo was taken by Denise Applewhite in 2009. Over-all book design is by Wanda España.

PREFACE

With no more family in Princeton in 2012, I decided to join two of my four young in Portland, Oregon. I returned to Princeton to pack things up and Dan Rubenstein kindly suggested a farewell luncheon with my colleagues. There I told a few anecdotes from my early days in the Biology Department and I was urged to write this up.

Some aspects of the history of biology at Princeton during my 65 years here overlap with the concerns of the whole university of which we were very much a part. I have already written about some of this in my autobiography (*Lives of a Biologist*) and I have directly lifted those sections and included them here. They are on the arrival of coeducation and the student revolution; they both affected us all. I have also included some matters concerning Robert MacArthur, and my becoming chairman.

I have been very fortunate to get the help of Henry Horn, my colleague of many years. He dug up the illustrations, including some of his own, and composed the captions. And as if that weren't enough, in two places where I had the facts wrong, he rewrote a bit in perfect John Bonner prose to cover the wound.

EARLY DAYS

When I arrived at Princeton I was given the use of an autoclave, which is a sterilizer, so essential in any microbiological work. It was the most beautiful, horizontal copper cylinder. But it was not perfect! Once the water above the gas jet began to boil one had to immediately turn the gas down so that it did not build up too much steam pressure and blow out the safety valve. Already then I was an absent minded professor in the making and inevitably the day came when I forgot to turn the gas down. It was at night and fortunately no one else was in the building, but when I got to the third floor of Guyot Hall, to my horror there was a dense bank of steam all up and down the hallway from the ceiling down to roughly my waist. I crawled back into the lab under the cloud bank on all fours and turned off the gas. This immediately allowed me to progress, for the Department bought me a modern autoclave with a timer that did everything automatically. More than beauty is needed in an autoclave and modern laboratory science was taking over. It was a transition point for all of laboratory biology and I was there for the before and after.

Right after I received my Ph.D. degree in 1947 I needed a job and I applied to three openings: Princeton, Johns Hopkins, and Amherst, and that is how I came to Princeton. Of course I had to give seminars, and the one at Princeton seemed to go well. I remember the next morning after breakfast sitting with Elmer Butler, the then chairman, and his phone rang. It was E. Newton Harvey, their most distinguished scientist, and generally believed to be the real power in the Department. But Butler kept saying, "Newton, I cannot discuss that with you now." I immediately thought that Harvey might be on my side, which happily, turned out to be the case. I was offered an assistant professorship at a salary of $4,000 a year which I eagerly accepted; Princeton seemed like a wonderful place for me, and I have not changed my mind 65 years later.

What is different today is that one does not usually start as an assistant professor if one counts all those intervening postdoctoral years, but no doubt the sudden glut of students at the time accounted, at least in part, for this. The Second World War was just over, to the great relief of the whole world. This produced free college education for all military veterans who were being demobilized in a vast steady stream. The government hatched the GI Bill to the joy of everyone. And today the salary seems incredibly modest, but that can be accounted for by inflation over the many years.

I have been asked to write down what it was like to join the Princeton Biology Department what now seems like many years ago. After brooding over the matter I decided it would be fun giving it a try. While my memory is still in moderate shape I do not have total recall, to put it mildly. I have been greatly helped by friends and family, who were there at least part of the time. When we arrived the oldest of our four offspring, Rebecca, was three years old and the next, Jonathan, was about to arrive.

I want to begin by saying something of the world of academia in 1947. Along with millions of others I had just been liberated from the Army. (Actually I was in the Army Air Force where I had worked in a laboratory on high altitude physiology.) As soon as I was free I made a bee line back to Harvard to complete my Ph.D. It was fortunate that I could do this in a matter of months because I finally managed an experiment that actually worked and carried me over the top. (I was able to show that cellular slime molds became multicellular by the aggregation of their amoebae to central collection points by chemical attraction. This notion of chemotaxis was not very popular at the time, but it gave me what I needed for a thesis.)

All veterans of the Second World War were eligible for support from the GI Bill of Rights and it had a great effect on education in the United States. It meant that all colleges and universities had a sudden surge in enrollment and needed to increase their teaching staff. I had finished my degree at just the right moment. It was imperative for me to get a job with my growing family, so I went on the market.

Another aspect of a new job that is different from today is that there were no starter grants. I was asked by Elmer Butler for a list of things I would need to get my research going and I wrote back with a simple list: Microscope, dissecting scope, autoclave, glassware, and other essentials. I was told they had the microscopes, which were excellent, and the autoclave (which as we have just seen was not quite adequate), and I could order my other needs after I arrived—the Department would foot the bill.

This was shortly before the days of the National Science Foundation; applying for grants there came only a few years later. I have told this often before, but on my first very modest grant I was asked to write an annual

FIGURE 1. The youthful John Bonner with his first Faculty microscope circa 1950 (photo courtesy of Princeton University).

report in the form of a letter, on my research progress. At the end of the first year I wrote that things had not gone too well; I had tried this, that, and the other, and nothing really worked. I got this wonderful fatherly letter back saying don't get discouraged—research often goes like that and he wished me better luck next year. Such an exchange of letters is part of the dim past: unfortunately neither investigators nor grant agents talk that way any more.

One great difference between now and then was the simplicity of our laboratory needs. This was not only the case for me with my work on cellular slime molds, but for others in the Department doing developmental biology mostly on amphibians in the laboratories of Elmer Butler and Gerhard Fankhauser, mammalian physiology (endocrinology) of Wilbur Swingle, cell physiology (and in particular bioluminescence) of Newton Harvey and his scientific offspring, Frank Johnson and Aurin Chase, more cellular physiology on red blood cells in the laboratory of Arthur Parpart, insect cytogenetics of Kenneth Cooper, plant physiology of William Jacobs. As far as laboratory techniques were concerned they were limited. A number, including myself, did histological sectioning. We all had microtomes and a collection of coveted dyes for the rituals involved. Elmer Butler even had his own preparator, Vincent Gregg, who was a master at making slides and often helped me with good advice. His position was paid for by the University; it was a perk for the professor of anatomy and embryology. The post vanished when Vincent retired. All those procedures of slide making go back to the nineteenth century, especially in Germany where experimental biology reigned supreme. I have been accused of being a nineteenth century biologist, and this was meant as an insult—someone who was behind the times. But it a true reflection of my work over the years; there were still many things to discover the old way to guide the new way, and that will probably always continue to be the case. That is possible because the questions the modern molecular biologists ask are often the same as the ones asked fifty years before, although the answers may be somewhat different: more penetrating and revealing in a different way. The danger is that sometimes the molecular gymnastics become so absorbing in themselves that the quest to solve the fundamental biological problem gets lost by the wayside.

Without grants, and even when the grants arrived, all laboratory equipment was purchased for one by Val Sylvester. He was a delightfully independent non-faculty member of the Department who definitely had a mind of his own. He would look at my list and would immediately say something like, "why on earth would you need twenty of those?" I had my first mini battle with him when I put in for some weighing paper; in those days I did all the media making, &c. myself. Val was about to retire and he had never heard of weighing paper; you just dumped everything on the metal pan, and then wiped it off for the next weighing, or snipped off a bit of newspaper. So, without telling me, he didn't order it! We had a hot argument and I finally persuaded him that weighing paper was the way to the future of biology.

FIGURE 2. Princeton Biology Faculty 1948-49. *Top Row*: William Jacobs, Gerhard Fankhauser, and Aurin Chase; ... *Middle Row*: John T. Bonner, Wilbur Swingle, Kenneth Cooper, Colin Pittendrigh, L. W. Silvester, and Frank Johnson; ... *Bottom Row*: Elmer Butler, C. F. W. McClure, Edwin Grant Conklin, Arthur Parpart, George Harrison Shull, Lewis Cary, and E. Newton Harvey (photo by Orren Jack Turner Studio, Princeton, NJ).

When I came to Princeton, biology was at a turning point. pH meters were considered an early step towards modernization although many of us still used crude pH indicator paper. Another change was in balances. I had been trained to use a Chainomatic balance to weigh out small quantities of a substance, a slow and cumbersome way, and now one just plunks that substance on a pan and the weight is instantly shown in big figures on the dial. Another trivial, but great step forward was the advent of plastic Petri dishes. They used to be made of Pyrex glass and that had some great disadvantages: they had to be washed, and to look at a culture with a dissecting scope the glass cover so distorted the image that it became almost impossible to see the result of an experiment. One could buy at great expense Czechoslovakian dishes that were optically flat, but they were made of soft glass and soon became very scratched. Then came plastic Petri dishes after I was at Princeton a few years. They were optical-ly perfect and it was the end of those dishwashing binges. Halleluiah!

Just at that time the ways of laboratory research were changing dra-matically, almost entirely due to the invention of new techniques, new tools. This did not occur overnight but over decades during the mid cen-tury. Much of it required big and expensive equipment. I think of the electron microscope as leading the way, followed much later by magical confocal microscopes, and innumerable other clever instruments that can reach into the hidden inner secrets of organisms.

In the middle of this last century biology made a number of great leaps forward. The most dramatic one that directly influenced so many fields, such as genetics, cell biology, developmental biology, physiology, and even evolution was James Watson and Francis Crick's discovery of the chemical nature of the gene and how it replicated. This was the birth of molecular biology and it generated many new instruments and tech-niques that allowed for a very rapid progress; the contributions of a vast number of workers in the field who continue their mass attack on funda-mental biological problems, still full steam ahead right into this century and the present day. Molecular biology may be the most dramatic, and indeed important thrust forward, but there are others, and as we shall see Princeton biology also played an important role in their advancement.

The University community is a very different place now than it was in 1947. Perhaps the biggest differences are all related to size. There are more students now, both undergraduate and graduate, although the numbers are still modest today compared to many other universities. The faculty has greatly expanded, not just because of the increase in students, but also because over the years Princeton became increasingly a research university, and while it clung to its reputation as a top teacher of under-graduates, it greatly expanded its graduate program with a big emphasis on research. I still have a catalogue from around the time I arrived. It is a slim volume that listed the administration, all the faculty, their home addresses and all the courses. The equivalent became a number of thicker volumes that dramatically illustrated how over the years Princeton has become a much larger place. (And now it is on an online digital version!) When I first came there were ten faculty members in the Biology De-partment—I was the 11th; now it has metamorphosed into two separate

Departments: Molecular Biology, and Ecology and Evolutionary Biology. Furthermore the boundaries of these new Departments of today are less sharp than the old ones and there is a growing interest in inter-departmental activities, both in research and teaching.

My first year overlapped with the last year of Dr. Cary who retired as an assistant professor: he taught parasitology to undergraduates. I do not know why he was never promoted; there were all sorts of scurrilous rumors, but I suspect it was because he did no research. My second year I inherited his wonderful office (which I inhabited for 64 years), but I did worry as soon as I moved in that I too might retire as an assistant professor!

It was not just the science that has changed, but the whole faculty community. The University League, made up of faculty wives, mothered all the newcomers and helped them become part of the community. However things were different before the war, and new faculty and their wives were then told very precisely what to do. They even had printed a small brochure to welcome and inform the new arrivals how to behave. I wish I still had a copy of that priceless document. Some of it went way back and even had directions on what to tell the maid, something that seemed quaint after the war. The world was much changed and all new assistant professors thought of themselves as now carrying out the duties of maids.

I feel as though my fellow newly arrived assistant professor, Colin Pittendrigh, and the Bonners made a significant contribution bringing in the new, post war changes to the Biology Department and the University community. We were not given the instructions on how to behave from the University League so we had no idea of what was to happen. On a specific Sunday afternoon after the semester began we would be called on by the chairman and his wife and we were to return the call two Sundays later.

We started our living in Princeton in what was called "the barracks." It had been the home for the accelerated naval cadet students who disappeared with the end of the war. The very simple units were immediately put to use to make living quarters for new young faculty and quite a few students. A big change had come over Princeton. Before the war there had been an inviolate rule that no undergraduate was allowed to be married. Now many of the returning students, as veterans, were not only returning to their studies, but they came with wives and often small

FIGURE 3. The "barracks" of the text were erected as "temporary housing" in 1948 on Princeton's former Polo Field. They have long housed graduate students as the Butler Apartments. They are currently under temporary reprieve from their latest scheduled demolition (2014 photo by Henry Horn).

children. The rules were quickly changed and many of our neighbors were undergraduates, but of a different breed. The hiatus of usually four years brought with it a new maturity; they and their families made good neighbors; they were undergraduates with gravitas. One undergraduate neighbor in a house just like ours (they were all pretty much the same) was the center of a great deal of attention (and gossipy talk). They had a magnificent Rolls Royce parked by their house; they had a nanny to take care of their small child, and a large German Sheppard that the gossip claimed was fed on steaks!

Much to our surprise one Sunday afternoon, we suddenly had callers; it was the chairman and his wife: Elmer and Eleanor Butler—they had a copy of the rules and we and all the other new faculty had not been so fortunate. We would have known that we would have been called on at a specific Sunday and that two weeks later we would call on them. The only problem was that when they arrived we were both in rather decrepit blue jeans and the kitchen sink was full of dishes. Ruth was quite up to the occasion. It was my job to make fascinating talk while Ruth magically subdued the sink and produced (presto) a pot of tea and some cookies and cakes, and joined us in no time in the living room. The Butlers were splendid and they helped us get a sprightly conversation going—there was no embarrassment anywhere, even though they were far more elegantly dressed than we. A copy of the rule book was found and it plainly said we should return the call two weeks later. We went quite dutifully dressed in our best (or was it our second best?), and the Butlers were wearing their blue jeans! Great laughter and a lovely tea. Eleanor had already washed the dishes before we came.

Colin Pittendrigh and his wife were less fortunate. He was the other assistant professor in biology who arrived the same time I did. They too were not forewarned so Colin, without thinking, yelled "come in" when the Butler's knocked on the door. But unfortunately Colin's delightful sister and a brilliant (and nice) graduate student from Columbia University were wrestling on the floor in their underwear having a tickling match. I think we were luckier. Those were the death throes of the University League system for welcoming new faculty.

My colleagues were an interesting, rather tight knit group. We met for faculty meetings around a big table and it was quite clear to me that there were a few large egos. Everyone seemed to get along, at least on the surface, but to a man no one was going to be pushed around. To my uninitiated mind many of the older faculty seemed remarkably conservative, but that conservatism was not on biological matters; it was entirely on social issues. Perhaps this, more than anything, made me feel I had joined a men's club. Their conservatism did not bother me; their biology was what interested me, and among them there was a lot of first rate biology, but of course that was not what was discussed at the meetings.

I was told to submit to the faculty before the next meeting an outline of a course to teach the next year. I wanted to give a course on the development of primitive organisms, which, just like vertebrates and sea urchins, had a development too. And their variety was enormously rich; it would include algae, fungi, protozoa and of course my beloved slime molds. I was totally unprepared when I came to the meeting when I was virulently attacked by Frank Johnson, one of my new colleagues. He said I was just going to teach the same things he taught in his microbiology course. Of course that was not true at all, but Frank did everything a bit differently and as a young neophyte I felt totally crushed. As I look back at this bit of theatre it fills me with pleasure to think before very long we became fast friends; I became an ally; not someone out to get him. The other more immediate sequel to this story is that the older members of the Department leapt to my defense and urged that we get together with our course outlines and Frank was soon convinced that I was not going to invade his territory.

I have always been fascinated by Frank; he was such a wonderfully complex person. In the beginning with others we shared the same secretary. Mary was a very nervous and shy person with very little self-confidence and it was soon obvious to me that she was terrified of Frank. Sometimes when I was there he would bring in a manuscript that needed typing, plunk it on her desk and say, "this has the highest priority. Stop everything you are working on, and do this first." The inevitable sequel was when he came storming back with the complaint that there were numerous errors and that it had to be retyped. Poor Mary. But I learned something that has kept me in good stead all these years. After one of these scenes I had a manuscript ready for typing. After some heavy thinking I brought it to her and said, "Mary, there is absolutely no hurry with this paper. Do it when it's good for you." Later that day she brought it to me all smiles. It was finished and virtually error free. Human nature bursts forth in many manifestations.

There was another thing about those faculty meetings that are different from the ones today. The majority of us smoked: usually a pipe, but also cigarettes and an occasional cigar. I remember after a bit it was like being in a thick fog; everyone was surrounded by it, but no one seemed to give it a thought. The smoke was the oil that kept the meeting running smoothly.

And now more about the faculty. As the reader may have gathered Elmer Butler was the Rock of Gibraltar and a splendid chairman. Soon after I got there he was replaced by Arthur Parpart who perhaps did not quite wear his new power with equal grace. He did have his battles with the staff whom he ordered about rather severely. Behind his back they referred to him as the "abominable NO man." The chairman's office soon became a matter of wonder. Arthur never threw away any papers but put them in high piles starting on the floor. It looked somewhat chaotic, but I soon found it was not. If I came and asked him about some past report or letter he would give the matter a moment of reflection and then lunge to one of these columns of stacked papers, take aim about a foot below the top and pull out the needed document. It was a remarkable performance; the world may seem disorderly, but there was perfect order in his brain.

The faculty members I became closest to in those early days were Kenneth and Ruth Cooper. Ruth was finishing her Columbia University doctoral studies in embryology at Princeton, and Kenneth had also done his graduate work there (that is where they met) in cytogenetics under Franz Schrader and his wife Sally; two exemplary biologists and two utterly delightful people, full of fun.

When I arrived in 1947 Ken was my senior by seven years had been there for some time and was an associate professor. Both he and Ruth took me under their wing and helped me launch my first teaching and academic duties. Ken in particular would make great—but kind—fun of me because of the inferiority of my graduate training: Harvard was totally backward compared to Columbia. The result was that I was trained all over again by Ken (with gentle nudges from Ruth) and learned some fascinating cytogenetics, a good deal of hard thinking about evolutionary biology, and even how to use a microscope! What kind of a graduate program did Harvard have!! The only place he did not prevail was in wasp taxonomy; he was rather disappointed I did not take up his passion.

At the time Arnold Toynbee was making a splash with his *A Study of History* in which he took a mega view of the rise and fall of civilizations. His ideas were widely discussed at the time and Kenneth and I got into a fierce argument concerning one of the theses of Toynbee's major work. I no longer remember anything about the substance of the argument, but I do remember that it went on for three days and that it was vigorous to say the least. Finally, the three of us were together, and Ruth in her calm way put her foot down. "This is the most ridiculous argument I have ever heard. Neither of you know what you are talking about: neither of you have even read the book!" That brought the discussion to an abrupt halt.

Somehow this memory encapsulates both of them in my mind. Ken, who loved to argue and was a master debater, while in Ruth's case her common sense intelligence came through and always prevailed. In later years, after they left Princeton, I would see them periodically and corresponded with them often and they did not change. They were a complementary couple and together they made a perfect match.

One thing I certainly will never forget. They decided to adopt a baby and after an interminably long wait the agency in New York called them up one day to tell them a very young son was waiting for them. They came to me in a great dither and practically asked me on their knees to come with them; I had so much experience with small tots. I was delighted and honored and of course immediately agreed to go. I was only sorry that my Ruth could not go too, but she was very fully occupied at home. And I was useful at a crucial moment. After admiring the baby, the agency worker came in the room and handed poor Ruth Cooper a clean diaper (the wonderful old cloth kind) and two huge safety pins and left! I have never had such a perfect chance to show off before or since then. "Ruth, stand back and watch this." All my skills which I had perfected over the recent years were put to use: Geoff had clearly been packaged by a pro. It was a touching moment for us all: no words, just a few hugs.

I have delayed talking about the Zeus (and Mrs. Zeus) of Mt. Olympus too long. Newton Harvey and his wife Ethel were formidable characters, but in very different ways. She published an early paper in which she showed that Hydra had an "organizer" very similar to what Spemann and Mangold discovered in amphibian embryos and for which Spemann was given the Nobel Prize, (but not shared by poor Hilde Mangold because she was killed before the Prize by an exploding gas tank in her kitchen.) What has always puzzled me is why was Ethel Harvey's extraordinary experiment, with all its insights into embryonic induction, never heralded? Was it because Hydra is not a vertebrate which automatically makes the work not comparable (which, of course, is nonsense)? My own theory is that Ethel herself failed to appreciate that her Hydra experiment was the first demonstration of embryonic induction. She was no shrinking violet, to put it mildly, and I think had she realized what she had done the world would soon have known of her great advance. She did another ingenious experiment in which she centrifuged sea urchin eggs in such a way that the nucleus was extruded, leaving a beginning embryo with no nuclei in any of its cells. The big question was how far would it develop without nuclei, and the answer was not far, but it did go through all its early cleavage stages so it could do some things without any nuclei, although not a

FIGURE 4. E. Newton Harvey and Ethel Harvey, "Zeus and Mrs. Zeus of Mount Olympus" circa 1950 (photographer not recorded).

lot. She fully appreciated the significance of this experiment and was well known for it.

She had a very strong character and felt it was quite alright to get young assistant professors to do work for her. After all they were young and should be happy to collaborate with a distinguished person. I got quite good at avoiding her demands, but I must admit, she won many of the battles. She tried her wiles on Kenneth Cooper, and the result was open war. He put an archery target on a locked door that connected their rooms and whenever life was more than he could bear he shot arrows into the huge straw target attached to her door.

Ethel Harvey was one of those many women who had the misfortune of being born at the wrong time. Today she would have been one of the leading scientists in the Department. And a happier person. Degrees, research accomplishments counted for little; one had to pay for being a women. I remember when Ethel asked me to ask Arthur Parpart, the chairman, if she should not be in the Departmental photograph; he was quite irritated with me for even asking. The answer was NO, and the reason was that she was a woman. Was I out of my mind? In today's Princeton she would have been much happier.

Newton Harvey was a very different person. He was a large man with a welcoming face and a way of conducting himself that made it clear, in the nicest, quiet way, that he was the one and only E. Newton Harvey. He always reminded me a bit of W. C. Fields: the way he looked, but especially the way he talked. His sentences were like Field's; they sort of drifted off on a high note. But make no mistake, he was E. Newton Harvey. His manner of speaking was easy to imitate and once one of his graduate students, who had perfected the art, accidentally answered him back in the perfect Harvey voice. It was not well received!

More than anyone else in the Department he had an international reputation. Younger scientists from Europe, in particular, would come to his lab for a year or two to learn Harvey's cell physiology. He was a pioneer bringing in biochemistry to understand how cells work. One of his most successful approaches was using bioluminescence as a way of getting some cells to tell the observer what was going on inside. And "living light" had a very general appeal to biologists and laymen alike.

Harvey loved to give Departmental cocktail parties and they were good parties. He got everyone to drink a little more than they intended; the making of a successful party. He would walk around the room with a huge cocktail shaker, all smiles and a good word for everyone, and while they were smiling or laughing, would top up their glass.

At one of these parties I was talking to Wilbur Swingle and a few others, and it was only after the alcohol haze had gone after the party that I realize Wilber was having me on. Wilbur was one of our senior professors and had great success in his research on the secretions of the adrenal glands for which he was well known. He taught the general biology course for many years; he followed in the footsteps of Dr. E. G. Conklin who taught it since its beginning. Swingle's lectures were much respected by the students. After a few refills from Harvey's mega cocktail shaker, Wilbur turned to me and said that he had just been teaching Ernst Haeckel's "biogenetic law" (ontogeny recapitulates phylogeny) and I looked suitably horrified for at that time it was in particular disrepute. I fell into his trap for he immediately said, "Oh I know, but it's so easy to teach." I was too hot under the collar to realize he was goading the young whippersnapper. Only long after did it dawn on me that Wilbur was giving me the business. The irony of this story is that Haeckel's idea has more or less come full circle and is now accepted as a valid, albeit a loose, generalization.

Gerhard Fankhauser was a bright star in our faculty, but he was the sort of person who blew his own horn so gently that only those who knew his work and his teaching appreciated his stellar qualities. He showed, in a beautiful set of studies, that newts with experimentally induced different cell sizes produced organs of normal proportions and shapes. So a newt made up of very few, but very large cells was indistinguishable in its overall size and morphology from one made up of many minute cells. (The amount of nuclear material was the same, only packaged in a few big, or many small cells. We still do not really understand how the shape and size of the animal is determined, but we know now, thanks to Gerhardt, that it is not controlled by cell size.) As for teaching he gave a course called simply "natural history" that was a gem. It may sound a bit old fashioned today, but what with the modern resurgence of ecology, it is a pity he still is not with us. It is not only for the timeless subject matter, but I always found it hard to believe that he was Swiss, for his English, both spoken in lecture and written in his published works, were enviable in their clarity and well-balanced sentences.

I went with Gerhard to a meeting of the Growth Society that took place at the University of Wisconsin in Madison and as soon as we got there he located a mutual friend who had a car. He wanted to drive to a town not too far off called New Glarus. In the middle of the 19th century there was a small emigration from "old" Glarus in Switzerland where crop failures had led to starvation. They prospered in Wisconsin and became recognized as an ideal, little Switzerland, something that made Gerhard feel it required a visit. When we got there he surprised us when he announced that we must go to the bar for information. I did not associate him with bars, but as we walked in he broke out with some question in Swiss-German and immediately got everyone's attention. They asked him his home town in Switzerland (it was a small town outside of Bern) and the bar tender rushed to the phone because one of their citizens came from the same town. He turned out to be a very comfortable looking man who sold cow insurance. His first question was what was his family name and when told the man said, "Was your father the town's family doctor?" The "yes" to this question produced a great exclamation: "He nearly killed me!" Apparently as a boy he had a terrible reaction to an injection he had received. This was followed by great laughter and approval that he had survived Dr. Fankhauser's ministrations.

Gerhard was greatly skilled at raising his delicate newts; town water did them in quickly. But there was a small spring that bubbled up through the floor in one of the rooms in the basement of Guyot Hall. He tried it and immediately discovered his baby newts thrived in it; Guyot was built over the purest of water. Unfortunately with progress the spring is now completely sealed over, which leads me to digress about Guyot Hall and its environs when I first came.

Guyot was built in 1910. It had a flat roof rather like a brick Victorian castle and when I first came it was forever leaking. There were endless work crews laying down tar and who knows what else so roof repair seemed to be endless activity. Guyot was finished at the time of Edwin Grant Conklin's arrival and half of the building was to house Conklin's new Biology Department—the other half was the Geology Department.

Once the roof problem was solved the building has served as an admirable home for biology. It was a very solid structure in all respects. It always fascinated me to remember that it was a very modern laboratory in 1910—the latest in scientific design with features that always seemed to me amazingly primitive. For instance when I came there were no hot water pipes for the whole building. At a later date small hot water heaters had been installed in the various labs and in a few offices. There were gas outlets everywhere for emergency illumination, even in the hallways and I believe they still may be functional today, at least they were when I last tested them.

For me what was much more interesting were the 1910 chemical hoods. They were beautiful wooden structures with glass doors and in their back walls they had good size vents, a lower one for air intake and an upper one for exhaust. Today in a modern hood that upper one would have an electric fan to move the air, but these ancient hoods had a gas jet! Once lit it would shoot warm air into the vent and cause the air to circulate by convection. It always reminded me of my lab course in organic

THE AMERICAN ARCHITECT

FIGURE 5. Guyot Hall's nearly finished front façade circa 1909. The Biology entrance is on the left; Geology, on the right. The capitals on the Biology and Geology sides are adorned with carvings like the insets, respectively extant and extinct creatures, some from the studio of Gutzon Borglum of Mount Rushmore fame. Note the skylights that contributed to roof leaks mentioned in the text. This view is now blocked by the growth of replacement trees; otherwise the main differences are that ground grade is now near the base of the lowest windows, and chases are visible between the towers on the left for the mid-Century chemical hoods mentioned in the text (*The American Architect* July 1909; insets are 2012 photos by Henry Horn).

chemistry where we had to distil ether. A built-in method of blowing up Guyot Hall!

My second discovery about the hoods came some years later. One afternoon my daughter Rebecca in her early teens, or possibly a bit younger, dropped in on her way home from school. Somewhere she acquired some incense sticks and she wanted to light them in the safety of my hood. I was very busy at the moment and said NO and bawled her out for being a bother to her poor, overburdened father. Somewhat later I rushed off to go the basement, four floors down and, like being struck

FIGURE 6. Bonner Family 1957. *Back Row*: Rebecca, John, Jonathan, and Ruth; ... *Front Row*: Jeremy and Andrew (photo by Willard Starks).

by lightening: I could suddenly smell incense in the hallway where there was a hood near by. It turned out that as soon as I left my lab, Marcia my sweet lab assistant, decided that I was being totally unreasonable (typical father behavior) and she and Rebecca rushed to the hood and ignited the incense. From it I learned that all the hood vents in the building were interconnected in a vast network, and that in no time the smell had traveled from the third floor to the basement. Today Guyot has nothing but the latest in chemical hoods, perfect for secretly burning incense.

Behind the building was a pretty pond and beyond that a rather decayed brick building called the Vivarium, now long since gone. When I first came it housed a number of animals for teaching and experimental work such as Wilbur Swingle's dogs; beagles that supplied his adrenal hormones. It always bothered me a bit because he cut their vocal cords to lessen the noise in the kennel, so they would greet you with this pathetic sound—more like an intake of breath than a bark. But who really knows how they felt about it. At the end of the Vivarium was a roomy greenhouse, full of tropical plants, but it had a curious neglected and decaying atmosphere. It has been replaced by a modern greenhouse on the splendid (and by comparison) huge Lewis Thomas laboratory building of Molecular Biology. There was nothing but green behind the Vivarium then. The distinguished geneticist G. H. Shull grew his plants there for his work in plant genetics. He had quite a collection of arrowheads and other Indian artifacts which turned up at the annual plowing.

A number of emeritus professors were about when I came, and George Harrison Shull was one of them. In fact it was from him that I inherited that beautiful, but troublesome autoclave that I mentioned earlier. Professor Shull was one of the two people who independently discovered hybrid vigor in breeding corn. The other was E. M. East who ended up at Harvard. Crossing two different pure strains is very likely to produce larger and more vigorous offspring. This hybrid vigor soon became important in agriculture; hybrid corn became enormously profitable for cattle feed, and indeed it still is today. (I cannot refrain from a personal note. When I took the basic genetics course as a undergraduate at Harvard it was taught by Professor E. M. East. He was a gentle, shy man, quite different from Professor Shull who always managed to let it be known that he was in the room. The tragic part of this story is that Professor East died in the middle of the course to our shock and universal regret.)

There were two G. H. Shulls wrapped up in one shell. There was the soft-spoken, gentle man who was genuinely kind and helpful, but there was always the possibility that in a flash he would transform into one of the furies and give a formidable display of outrage. Something one had said or done was on his black list and it was not all that easy to earn a pardon, but when it came it was equally abrupt. He apparently had one of those altercations with his bank and as a result he kept his money in one of the locked cabinets in his old lab, just across the hall from my office. With considerable regularity he would enter his old territory, look furtively in all directions, unlock the cabinet remove some ready cash, lock up, and flee. I never saw him make a deposit, but I assumed he got the cash directly from the treasurer's office: no incompetent middle-man for him!

When I arrived in Princeton the Moffett Laboratories did not exist; Guyot ended rather abruptly with two teaching laboratories on the main floor. One of them was also used for weekly seminars which have left a variety of memories with me. For instance I remember when the distinguished spider and insect authority, Professor Alexander Petrunkevitch came from Yale to give a fascinating talk on spider behavior: good old-fashioned natural history. The audience was rather thin, probably because the subject was not a central issue at the time, but there were two children who looked as though they were about to be teenagers in the front who made the day for the speaker and the rest of the audience. They had obviously read a lot about spiders, more than anyone else in the audience, and seemed to know all about Professor Petrunkevitch's work. He soon took the measure of his audience and proceeded by giving the lecture directly to the very young. It was a great show, but unlike any lecture I had been to before or after. The children would sometimes challenge the speaker who not only parried with the skill of an old pro, but with genuine delight in his youthful audience. They were the offspring of Doreen and Lyman Spitzer. Lyman was Princeton's stellar astrophysicist; a remarkable family as was obvious to us all.

The chairs in this room were made of bare wood and incredibly hard and uncomfortable. Newton Harvey always arrived with an inflatable rubber ring to make attendance bearable. My first year I had the surprise of my life on the first warm day. Needless to say this was years before air conditioning and it got very hot in the room. After the clapping we all stood up to leave and the chairs came with us! The wooden seats were generously coated with varnish that had melted in the great heat. That is a problem that happily has vanished into the mists of the past.

On the walls of those laboratory rooms hung large lithographs of various animals, vertebrate and invertebrate that were partially dissected. They were from the teachings of Thomas Henry Huxley who did so much bring the science of living things to the classroom and to the general public. He was Darwin's friend and defender of his natural selection in a period when it was much resisted. He also was a great teacher and a spirited advocate of having the organisms themselves studied and dissected in the laboratory; lectures alone were not enough. I have always assumed that those lithographs in Guyot got to Princeton through the influence of two gifted students. They were William Berryman Scott and Henry Fairfield Osborn both of the class of 1887. After graduation they crossed the Atlantic and took some courses in England and then moved on to Germany. One of the British courses was with Professor Huxley at the School of Mines. It was a famous course and its influence definitely reached Princeton, lithographs and all. This undoubtedly was the root of why we started as a Biology Department and not separate Botany and Zoology which was the usual thing in those days. For a few years, in the early 1900's, William Berryman Scott was chairman of both Geology (which was his field) and the nascent Biology Department, something that would no longer be possible today!

The best known emeritus professor, who was very much still there when I came, was Edwin Grant Conklin. He had been brought to Princeton in the early 1900's by the then President of Princeton, Woodrow Wilson, to essentially start a Biology Department. This involved a new building—Guyot Hall (shared with Geology)—and building a faculty. So everything that greeted me when I first came was the direct result of Dr. Conklin's endeavors. I was a great fan of his embryological studies of mosaic development of certain marine invertebrates. He and Ross Harrison at Yale, and a few others, put America at the forefront of developmental biology of the time. Suddenly America was placed beside Germany at the leading edge of developmental biology, and Princeton was right on that edge. (Conklin started his breakthrough study at Johns Hopkins University and the Woods Hole Marine Biological Laboratory, and later at the University of Pennsylvania, but he brought his considerable aura with him to Princeton.)

It seems to me as I look back on these older professors that many of them, both at Princeton and at Woods Hole, were imposing personalities. They made their presence felt. To me, the callow youngster, they were the inhabitants of Mount Olympus. This was certainly true of my feelings towards Dr. Conklin. He never said a cross word to me, yet I had heard he had a most impressive temper when he lost it. And it had nothing to do with my awe; perhaps that had more to do with my youth and inexperience in the world. I had the same feeling of respect for R. G. Harrison, E. B. Wilson, T. H. Morgan, and others, all of whom were in Woods Hole in the summertime. And in all ways they were a heterogeneous lot.

For instance, there was no more modest and gentle soul than Professor Harrison; the opposite pole from Professor Morgan who was very outgoing, yet to me it seemed only right that they should share occupancy on Mount Olympus. It was their great contributions to science and not their personalities that made them so exalted in my mind. For a great variety of reasons I think younger generations view their elders differently today. I am not complaining, but the whole world of biology (and other sciences) has undergone multiple transformations with far reaching effects on all our perceptions. This is in part due to the tsunami that is molecular biology, but also due to the not unrelated vast increase in the number of biologists and the number of new journals that have come into bloom. A journal article used to often have a single author and occasionally two or more. I do not know what is the world record, but we have all seen papers with 30 to 70 authors from institutions all over the globe. Mount Olympus is no longer a country inn but a Grand Central Station.

The first thing that happened to the new assistant professors was to be put on a University committee. I was assigned to the Library Committee; the Firestone Library had just been completed and the books were so crammed in Chancellor Green that many were stacked on the floor. They all had to be transferred to Firestone. Let me insert a comment here. Even when the rotunda there looked like the attic of chaos with its masses of books all over the place, one could plainly see it was an architectural gem with all its alcoves, and its stained glass was quite splendid. Fortunately this was appreciated by the University architects and today Chancellor Green looks more beautiful that ever; all the clutter is gone.

The other committee I was put on was the Scheduling Committee that was responsible for assigning the teaching hours for the various courses. I had not realized the complexity of the problem. Courses, especially required courses for many fields, such as organic chemistry for biology majors, could not be given at an arbitrary time. John Tukey, then a lowly assistant professor was at that time the human calculating machine and therefore the heart of the scheduling committee. The rest of the committee did nothing but gasp with admiration at John's performance. And a performance it was: with drama. John was rather a large man, and he insisted on lying down horizontally. And when he was quite comfortably settled, the Registrar would then start reading off the long list of courses and John pronounced for each one, for example, "History 207" and after a dramatic pause with minor groans he would say "Monday, Wednesday and Friday at 11 AM." The pronouncements were the final word. I had the feeling that the whole performance would stand up well to that of the Oracle of Delphi. Now a computer can do the same thing with equal reliability, but with less panache.

Princeton, and indeed similar institutions throughout the United States, were bastions of quietly suppressed racial prejudice. The world was changing, but by mid century Princeton was still lagging behind. Some of the older professors were quite worried about the admission of Jews and they would pore over the graduate student applicant's photograph for signs of Semitic features. (In recent years the requirement of a photograph on an application has been dropped.) It is hard to believe that we had that dark past. In biology the turning point came when we did the unthinkable: we hired a splendid cell biologist who happened to be Jewish. It helped that Lenny Rebhun, over and above being a first rate teacher, doing first rate research, was a cheerful man of great charm; we all felt honored by our new colleague. Since then there have been many others both in the Department and the University—even a Jewish president (Harold Shapiro) for some years. Many universities were slower to respond to what was happening many places elsewhere in the world but we got there eventually.

What I want to do now is go to specific periods and events that were important to me during my years at Princeton. The ones I have chosen are (1) my becoming chairman, (2) the no-holds-barred struggle between biochemistry and molecular biology and the rest of biology; where lay the future? (3) And how ecology and evolution came into bloom. (4) And we were all deeply affected by the arrival of coeducation in 1969, and (5) the student revolution around 1970. (6) I will close with a discussion of how biology itself has changed over the years, something that was always knocking at my door throughout my career. The changes, the progress, have been phenomenal and Princeton was right there all the way. An amazing voyage.

MY BECOMING CHAIRMAN

I came in for a big jolt in the spring of 1965. Bob Goheen, the President of the University, called up to say he was coming down to see me right away. This put me in a fluster, but along with everybody else, I was a great fan of his, so I was glad to see him; but why was I not summoned to his office? The moment he came, he explained his mission: he wanted me to be the new Chairman of the Department. I instantly had two thoughts race through my head: I didn't want to do it, but I was going to say yes, both of which turned out to be true.

I was to take over the next academic year. My predecessor, Arthur Parpart, was about to leave for the summer for the Marine Biological Laboratory in Woods Hole, Mass, but he promised he would tell me all when he returned, including the details of the finances of the Department, which made me especially nervous because I knew my severe limitations as a bookkeeper. The awful thing was that poor Arthur had a massive heart attack just before he was to return and died instantly. After the funeral and the memorial service (which was most moving because there were no eulogies, just a Bach solo piece for the cello, played by a gifted colleague—music does things that words cannot begin to manage), I tried to gather in the reins.

The first attack came from the Treasurer of the University, a splendid person named Ricardo Mestres, who told me that the Biology Department was paying the secretaries and the laboratory assistants too little, and that he could not let me see the Department accounts. They were being fed into a computer for the first time and no one had them! I proceeded to make a whole series of wild guesses, because Arthur had left nothing behind either—all the figures had been in his head. I fixed the salaries and flew blind all year. At the end of the year I had a meeting with Dick Mestres and all his accountants and associates. Dick liked to make these meetings solemn to make one feel one had done everything wrong, but what he told me was so outrageous that I came as close to blowing up as I am capable. He started the meeting by saying that I was paying the secretaries and the assistants too much, and I had been so economical that there was a surplus of money in the various accounts that the University would take back. The blood rushed to my vocal cords and I said,

FIGURE 7. Faculty mentioned in the text circa 1970 (Biology unless otherwise noted). *Top Row:* John T. Bonner, Elmer Butler, Ted Cox, Gerhard Fankhauser, and Henry Horn; ... *Second Row:* William Jacobs, Frank Johnson, Marc Kirschner (Biochemistry), Egbert Leigh, and Arnold Levine (Biochemistry); ... *Third Row:* Robert MacArthur, Arthur Pardee (Biochemistry), Colin Pittendrigh, and Lenny Rebhun; ... *Fourth Row:* Bill Bowen (Economics & University President), Bob Goheen (Classics & University President), Harry Hess (Geology), Lyman Spitzer (Astrophysics), and John Tukey (Statistics); (photos from Princeton University Faculty 1967-8, 1969-70, and 1972-3 face-books).

"You told me to do that for the secretaries, and your office helped me get the figures right. The only reason there is money in the accounts was that you still have all our figures stuck in your goddamn computer, so we had to deny ourselves many things we badly needed until we found out how much money we had." I realized a long time afterwards that Dick was just testing, and I must have passed, because we became good friends and I got what we wanted.

I learned many things that year that were revelations about human nature. When I started, an old friend who had the same job for a few years at the University of Edinburgh wrote to me that I should put up a sign on my office door saying, "Departmental Chaplain." Slowly my distaste for the job began to recede a bit. It was partly because I was able to keep my research going without any serious lapse, and partly because I continued the same amount of teaching as before. There were a few small rewards in terms of things accomplished, things that worked. For instance, I had an older colleague, the very same Frank Johnson mentioned earlier who had caused me some grief in earlier days, before we became good friends. He had a mixture of aggression against the world—and especially authority—and a splendid sense of humor. One day he called me to say he wanted to see me about a matter that had been bothering him. I told him to stay where he was, and I would be right down. He started off in his aggressive vein, "John, you're the chairman, you're the authority here, and there is something that is your responsibility." When I asked what it was he said, "People just aren't flushing the urinals in the building, and it's disgusting." I told him that would require some thought, and that I would come to see him the next day. When I returned, I told him I had the perfect solution: I was going to install a huge switch right over his desk that was connected to each urinal in the building, and whenever he felt the urge, he could pull the switch and they would all flush at once. There was a tense moment of silence, and then he burst out laughing. This shows why I occasionally had a feeling of accomplishment, but perhaps that was my greatest moment in my years as chairman.

There were more such surprises. One day I was sitting in the chairman's office when my colleague Bob Allen burst in, obviously in a state. He was a very tall man with a strong personality. He was the sort of person who tended to demand rather than ask, as he towered over one. I asked him what was wrong; I could tell immediately his equilibrium had received a knock. First I got the usual bit, "You're the chairman; you've got to do something!" I replied he'd have to tell me more before I could act. Then he explained that his research associate and his technician were "in his dark room screwing" and that I had to come immediately to stop it. I explained that there were many odd things a chairman had to do, but this was not one of them. If he felt strongly about the sanctity of his dark room, then he, and only he, must deal with it. At least, so far as I know, there were no new babies in the Department, just a "now I've seen every-thing" expression on the chairman's face.

Since I am on the subject of sex there was another new experience for me. One day I was sitting in the Departmental office and suddenly all the secretaries streamed in, single file, led by the Departmental secretary, a very forceful and determined woman. They demanded that I fire the new janitor! It turned out that the iron steps on the ancient staircase to the subbasement were designed in a lace-like pattern; they were the ulti-mate in "see through." The janitor would position himself under the stairs where he had a good view up the ladies' skirts. In those days all the wom-en wore skirts; slacks were unthinkable. So I had to fire the man; my first encounter with the unpleasant necessity of firing someone. And in the subbasement under each step was filled with some sort of solid, opaque material that is still there today and now all the women wear slacks. An-other win for Grace, the Departmental secretary: soon the Department would not need me at all!

The principle janitor was quite a wonderful person who made a point of not bothering, and only helping people and was admired accordingly. Bill Reap was the perfect senior janitor. Mostly he was not in evidence, but he did have his daily moment with the faculty. Those were in days long before our mail room and Bill would personally go to every faculty member's office. If one were at one's desk he did not hand you the whole pile, but gave them to you one letter at a time with a comment for each. "This one might be good; it might have a check in it; this is just junk; here is a fat one, probably from a friend; I bet this is a bill. You did not do too badly today! More real letters than junk."

I have another memory of those early years: my friends who were quite aware that I suddenly had a lot more to do and wanted to be helpful. I wrote out all my letters long hand on lined yellow paper and Grace, the departmental secretary, edited them so that they conformed to what she learned in secretarial school some time back (spelling, punctuation, &c.) For instance, I tried to persuade her that it was no longer necessary to put an apostrophe before 'phone. I worked quite hard on this, but no letter typed by her went out naked without an apostrophe.

Well, I was persuaded to buy the latest hand-held tape recorder so I could catch up on my correspondence in the evenings by dictating. It seemed like a grand idea, but in fact it ended up a disaster. My study in that wonderful old University house was right off the living room, separated by a very thin wooden door. The first evening I started dictating I suddenly realized there was giggling in the next room by the door. My wife Ruth and daughter Rebecca had found what I had to say, and how I said it very funny. And here I thought I was encapsulated in privacy when in fact I was making a public exhibition of myself, and a very inhibiting one. I gave the recorder away!

There are many, unrelated but interesting things that happened in those early days. One of them was how we acquired Eno Hall when the Psychology Department (and later, for a brief interval, the Mathematics Department) moved out to larger quarters. My phone rang and it was Doug Brown, the wonderful Dean of the Faculty. I was a great fan of his because he was so especially helpful to me in the beginning when Arthur Parpart was no longer there to tell me what to do. He said, "Could Biology use Eno Hall?" I said we certainly could; we were growing and were just about out of space. We chatted a bit and I gathered he had a meeting with Harry Hess, the chairman of Geology who told Doug that is was Geology's expectation to inherit Eno which is adjacent to the Geology end of Guyot. I don't know how Harry put it, but he infuriated Doug who decided to give it to us; he was not about to be talked to like that even though—or perhaps because—Harry was an admiral in the Navy Reserve. We made a big step forward without doing anything but answering the telephone. That is the way to run a department! And we really put Eno to good use because before long it housed our budding spurt of growth in ecology and evolution, led by Robert MacArthur. More on that shortly.

George Packer Berry was a loyal Princeton alumnus. After a distinguished career he had just retired as Dean of the Harvard Medical School and came to Princeton to live peacefully. Bob Goheen, the President, provided an office for him in Nassau Hall and he became an informal advisor to Bob. Being an emeritus bacteriologist his first concern was the Biology

FIGURE 8. Liberation of Eno Hall. To honor Biology's acquisition of Eno Hall, a pseudo-military occupation was organized by unspecified folks from the following list (graduate students unless otherwise noted): *Window:* Walt Ogston and Bill Schaffer; ... *Ground:* Henry Horn (Faculty), Ernest H. Williams, David Griffiths (Postdoctoral Fellow), Gary Sprules, Brad Lister, and Carl Wynter (photo by self-timer).

Department and its new, very green, chairman. Later I want to describe the "molecular wars," in which he was very much involved, but first something about him and our pleasant, if somewhat odd, relationship.

George was a big man who, even at his retirement age, was an incredible mass of energy and a charmer at the same time. He was never still: he was on a slow boil at the back burner and would leap to the front burner in a flash. He once told me that he got a call from his President, James Conant, when he was at Harvard, to drop everything and join him for dinner with some distinguished visitors. George rushed home to change, undressed and started putting on his pajamas. Too many burners on at once. He arrived at the dinner, suitably dressed and on time!

He immediately perceived that I did not really relish being chairman. I wanted to spend more time in the lab. I would do it with the best of my ability, but that was not enough—he wanted me to enjoy it, and he came very close to succeeding. He urged me to think out what I wanted to happen in the Department, and then think out what was the best way to have it happen. He thought of it as sort of war games (although when I became surrounded by the molecular problem, as we shall see, they were not games any more, but the real thing).

THE MOLECULAR WARS

In the middle of the century the arrival of molecular biology, which descended from biochemistry, was the beginning of a mega revolution, not just in Princeton, but in the whole world. We suddenly knew the chemical nature and the structure of the gene, the fantastic discoveries of James Watson and Francis Crick. These were without question the most important advances in 20th century biology. It is hard to believe that when I came to Princeton no one knew the chemical nature of the gene. In one bold coup Watson and Crick showed both that it was a nucleic acid and how it could replicate itself, so essential for gene-like activities. Then immediately there was a rush of a number of workers who showed how these nucleic acid genes produced specific proteins.

This was such a stunning advance that it gripped the world of science; it opened up great new vistas for every branch of biology from cell biochemistry to evolution. There was also a mindless after-thought soaring from this exalted position: all of biology must now be subsumed under molecular biology. All the people working on any other form of biology were barking up the wrong tree; that tree was all dead wood. The traditional fields such as developmental biology, animal behavior, plant physiology, microbiology, and so forth, unless they were being investigated using the new founded molecular techniques, were pointless hangovers from the past that would not survive.

Today we have managed a more balanced and sensible view where we do not throw out the old but add to it. By and large the big questions remain the same and some of them are ideally suited to be attacked at the molecular level. And this is being done with increasing success. For instance, developmental biology has made great advances with the new molecular techniques where we can delve deeper and deeper—at an increasing micro level—into the causes of the great multitude of developmental steps. It has even been very important in the study of evolution: by analyzing changes in the genetic makeup of a group of organisms it is now possible to establish very accurate phylogenetic trees so one can chart who is most closely related to whom, and not only when they first arose in the tree, but in some cases the actual time in geological history when they first arose. All these achievements are exciting advances, which in many cases have revolutionized our knowledge.

That is where we are now, but as I indicated above, in the first wave of post Watson-Crick excitement there was a more radical view, one with which I became all too familiar. The idea was that all the rest of biology, the non-molecular, old-fashioned biology was dead wood, suffused with dry rot. It was the past and all of it could be thrown in the dumpster. Molecular biology was the only biology. At the time I didn't understand or have a clue that anyone thought this way, because I kept asking myself how could I use molecular techniques in my work with slime molds to advantage and some of these applications produced very interesting results. (Of course the work was done in collaboration with molecular colleagues.)

Just before I became chairman, Princeton decided to hire some bio-chemists, something that had been embarrassingly absent for many years. Newton Harvey had been urging the department to go this way for ages, but it never seemed to happen, and now suddenly it did, but by then Harvey had retired. The chemists wanted to be part of the expansion too, so a pro-gram was created in which half of the new biochemists were to be housed in Chemistry and half in Biology. This was neither a wise or happy decision but it seemed to please all those who wanted a piece of the pie. The good part of the arrangement was that Arthur Pardee was recruited to run the new program and he was very effective in keeping both Chemistry and Biology happy. He had a strong personality and impeccable scientific credentials. And when I became chairman shortly after he arrived his presence was a great boon to me, both in scientific matters (we had a joint doctoral student) and in administrative matters. We became good friends and he was forever telling me that I made a lousy chairman because I was not tough enough.

Unfortunately, after a few years, Art decided he wanted to get back to the laboratory and some of the other senior members of the program took over the running of it. That was the beginning of the molecular wars. There now was a sharp line in the sand between Biology and Biochemis-try on two levels: administrative and intellectual. I would constantly be involved in petty fights: we had half a secretary more than Biochemistry and it was up to us to make up the difference. I must say the President and the Provost were wonderfully supportive and solved the problem for me so I did not have to saw a secretary in half. However, there was a constant struggle, both unsympathetic and aggressive, to see that we did not have anything more than they had. I was being harassed.

On the intellectual level it took me a shockingly long time to understand the problem: we were no longer biology; they were modern biology, and the sooner we just disappeared—went extinct—the better. It was so foreign to everything I believed that it had to be rammed home to me; my naïveté seemed to have no limit. I had thought that George Berry would be my savior in this struggle, but unfortunately he, along with the rest of us, failed. He made efforts trying to get the two sides to at least show some civility to the other side, but that never happened. He pushed for sharing some of our named professorial chairs to biochemistry that was no doubt as it should be, but the senior biochemists responded by asking (demanding) more. All the efforts to bring harmony were rebuffed so that even the emeritus Dean of the Harvard medical school's magic wand and cheerful reasonableness could not bring warmth and light to our two departments.

The next big step occurred when biochemistry was transformed into Molecular Biology and Arnold Levine returned to chair the new Department. Things definitely improved although many of the old tensions were still there. A few rooms on the borderline between Guyot Hall and their Moffett Laboratories were destined to become part of Molecular Biology. The business manager of Molecular Biology decided that they should have those rooms right away and moved all the books and signs of habitation into a closet on the beginning of a weekend. This included one of the rooms of Marty Kreitman, who was with us then, and it turned out, had a formidable temper. The Provost and some of the Deans came down to see the scene of the crime and calmed every one down, but the damage had been done.

This happened just after I had retired as chairman, but it set up waves that crossed to me over the Atlantic. The deans were sick of the war and decided peace should be declared. I was on sabbatical in Edinburgh, Scotland so the University flew me back for a few days to help pour oil over the situation. Neither I, nor George Berry, nor the administration could make Molecular Biology budge a millimeter: cooperation was out of the question; they were the future and they would lead unmolested.

After my initial 12 years as chairman I was asked twice to come back for a year because everything in the life sciences continued to be unsettled. It took some time to get to the happy equilibrium of today. Molecular Biology had recruited some very strong younger people but

FIGURE 9. Biology, including Molecular Biology, Faculty mentioned in the text, 1980's. *Top Row:* John T. Bonner, Ted Cox, Jim Gould, and Peter Grant; ... *Middle Row:* Henry Horn, William Jacobs, Marty Kreitman, and Arnold Levine; ... *Bottom Row:* Bob May, Dan Rubenstein, and John Terborgh (photos from Biology at Princeton University 1986 brochure to attract potential graduate students, plus Terborgh from Princeton University Faculty 1977-8 face-book).

unfortunately they did not all stay. One was Marc Kirschner who seemed to me particularly gifted and has gone on to a stellar career, ending up at Harvard. Our President, then Bill Bowen, was particularly concerned that Princeton was losing him and asked him to come to his house one evening to try and persuade him to stay. The next morning Marc called me and I had the most riveting conversation with him. He said that he had done me a good turn, and indeed he had in more ways than one. He told the President that the University did not seem to realize that the other part of biology, the non-molecular part, was receiving wide recognition in the outside world and that that is where Princeton's strength, both existing and potential, lay. The President was amazed because his informants had told him just the opposite. Marc said we had been dumped on relentlessly and that it was time for me to blow our horn. This was a big signal that we were at a turning point.

For me the ultimate turning point came a bit later. I got a call from President Bill Bowen's office asking if I could meet with him the first thing the next day. I had no clue why, but I was definitely interested and certainly not prepared for what he had to say. He was already in his office when I arrived and I thought he seemed a bit nervous. Here is the gist of his atomic bomb of a message. He had come to realize that all was not well in the biochemical-molecular part of the University and that something had to be done. He had not done anything yet but was going to do it that very day. The reason he wanted to see me first was that he realized how much misery I had suffered as a result of all the problems these last few years and he felt I deserved this consideration. As of this very day he will be relieving the two senior biochemists of all their duties, especially their administrative duties, and they will be the sole members of the Department of Biochemistry and that I will be their acting chairman! It was a stopgap measure that calmed the turbulence and was the beginning of the constructive peace that is still with us today.

I do not know all the forces that led to this mega decision, but I gathered that Princeton's problems were so well known all over the country and the more scientifically minded trustees knew all about it. Bill Bowen felt it was time to clean out the Augean stables.

(I showed my colleague, Ted Cox, who was one of the younger biochemists at the time, what I had written above and I was stunned to learn there were all sorts of machinations going on that I had never been told. It turns out that the need of a major shake-up was being busily looked into and the Dean asked Ted Cox to meet with the two senior biochemists to discuss the possibility of a mini department that Ted would chair. The talks disintegrated and the matter must have been put in the President's hands possibly leading to his call to me.)

I suppose the molecular wars left their scars. I remember during a particularly intense period, driving home one lovely spring afternoon during an exceptionally busy period I was driving through the countryside near Princeton and passed a farm that sold grass sod for making instant lawns. In the middle of a beautiful sweep of grass was a big sign saying: CULTIVATED SOD. I suddenly had an insane desire to come out in the middle of the night, steal it, and arrange it outside the door of the chairman's office. Of course, I never did, but somehow the thought greatly helped relieve some of the tension that was rising in me.

COEDUCATION

During roughly the same period Princeton itself underwent two unrelated revolutions: coeducation and the student rebellion. Of course these eruptions were not just taking place in our University, but all over the country, and many places abroad. Since its founding in the eighteenth century Princeton had been an all male institution and this was very much the case in 1947 when I arrived. The only women were the secretaries and it took some time after the beginning of coeducation before women had the same status as men, but they are closer to it now, and for the last 10 years our esteemed President has been a woman, Shirley Tilghman.

The big event occurred during the reign of Bob Goheen, and it was not an easy step. I was on some committee that met with the trustees and as I entered the room for the meeting Bob said to me he wanted me to meet one of the trustees. I forget his name but Bob put his arm around his shoulder saying, "This man caused me more trouble about our becoming coeducational, but I've just learned that his granddaughter has arrived in the freshmen class." Laughter and smiles all around, especially from Granddad. The group most resistant to coeducation were many of the all male alumni. The whole ambiance of the University will go down the drain. Educating women was a good thing in principle but not at Princeton—it should be done somewhere else. The letters to the editor in the Alumni Weekly were remarkably indignant, but today that seems like a long gone era.

In the beginning we were all a bit confused and uncertain. I always felt sorry for the first women to arrive, for the boys had no idea how to behave. I suddenly had women in my basic biology course, and the only effect I can remember was in the beginning going into the Men's room before the lecture to be sure my fly was zipped. It took no more than two years for all of us to find it difficult to believe that we had once been all male—everything seemed so normal now that we had what was quaintly called, "sex blind admission."

When a women's studies program started I was appointed to its committee. It was a very interesting, and indeed an eye-opening experience for me. Furthermore I felt quite useful because I plugged hard for having some biology taught besides the history, the literature, and the other relevant disciplines. The newly appointed director was an anthropologist and she embraced the notion with enthusiasm. This is one of the reasons Princeton had a strong program right from the beginning. My only problem was keeping the distinction clear in my mind between the meanings of gender and sex—it was rather like trying to remember the meaning of dialectical materialism.

At the very beginning there was a certain amount of aggressive behavior, but that soon dissipated into friendly joshing. Let me give some examples that I remember. One was when a student and I arrived at the front door of Guyot Hall at the same time in the morning going to work. I lunged for the door to open it for her and she gave me a furious look saying "those days are over: You go first." It was too early in the day for such radical thinking and with equal fury I said I've been opening doors for women for forty years and I 'm not about to change this morning. I prevailed, but no one was smiling.

Lecturing on general biology to a large class was something that I enjoyed. I did it for many years. The students were responsive and I seemed to keep control of that sea of faces, almost always friendly and attentive. Sometimes they would needle me; I remember one young woman who sat in the middle of the room, and whenever my eye would fall on her she instantly produced a large pink balloon with her bubble gum. Her timing was perfect and it always came as a surprise and produced a glitch in whatever I was saying.

I particularly remember one student who had been in the Israeli Army. She was a vital presence with flashing dark eyes. She sat in the front row and during the lecture at least once she would wave her hand and cite some exception to what I just said. I would say she was quite right but I had not mentioned it because it was not relevant to the point I was making. This went on for a few weeks and then one day the inevitable question came and I said this time you are wrong: I was basing my point on more recent work. Silence. Then the whole class burst out in applause. The old boy finally won a point.

Some time later at the end of my lectures a young woman came up to the podium and said that she and some friends would like to take me to lunch; the class ended at noon. I said I'd be delighted and they would pick me up after the Wednesday lecture. They were four and all splendid looking. What I had not realized was that I was still on the podium, some inches off the floor. I gathered my notes and stepped down and it was a bit of a shock to find that all four towered over me. They were all on the varsity crew. Lunch could not have been more fun; full of laughter. This was the new Princeton—more like the outside world. Everyone felt comfortable. The ambiance is back: it is a little different; perhaps even considerably better.

A closing recollection. The printed rules of the faculty naturally said "he" throughout, but now there were to be women faculty as well. At one of the monthly faculty meetings in Nassau Hall it was moved that every "he" should be replaced by "he or she." The English Department faculty were there in a block, and in a fury to vote against it. We would be wrecking the Queen's English! After a long debate a man got up and said in a gentle voice that he thought we had been so mean to women for so very long, he would recommend that every "he" be changed to "she" which would satisfy the problems of the English Department. The best moment in all my years going to those faculty meetings.

THE STUDENT REVOLUTION

The 1960's and 1970's were an extraordinary time of change for young people. Largely because of the Vietnam War—although perhaps that was just a trigger that released something that was going to happen in any event—the young staged an impressive revolution. The students at the University, as did the students everywhere, began to assert themselves. It was suddenly as though there was an instant formation of a powerful union, and that union was an irresistible force that could impose its will on the older generations, which were in shock and in disarray. The outward signs were most obvious in the male students: no neckties and long hair; and for both sexes, the universal blue jeans, free use of four letter words, the disappearance of chastity, and the appearance of drugs. At one point the students demonstrated around a building on the campus where secret military work was being carried out. A next-door neighbor told me that at dawn, before the crowd gathered, a small, undeveloped young girl of about fourteen was painting on the wall of the building, "Fuck the Bourgeoisie." At the University requests became demands, and the radical element among the students took over the undergraduate newspaper and would attack the University officials and faculty—any authority—in the most virulent fashion, a virulence that has only been matched by the far right in recent years. The barricades were finally stormed when our Government bombed Cambodia—that was the last straw (although there were quite a few last straws, such as killing the students who were demonstrating at Kent State University). The outcry was deafening and the whole University became convulsed. Besides student demonstrations there was a series of marathon faculty meetings that were wisely broadcast on the student radio station so that all could know that we were not conspiring (and the spouses could follow the drama and know when we were coming home for dinner). Bob Goheen, our President, showed remarkable decency, patience and enviable steel: his role was crucial in averting any serious trouble, or any regrettable turns.

Two of the undergraduates in the Biology Department came to see me to tell me that they had called a meeting of all the students, graduate and undergraduate, all the laboratory assistants, secretaries, janitors, and faculty in the Biology Department for the next morning. I told them they could not have it then; that would interfere with classes. They informed me very politely they were quite aware of that, and they hoped I would come! Of course we all did, and our big lecture hall was absolutely jammed—I do not think there was a soul anywhere else in the building. One of the students started the meeting, saying that it was held because of the outrage of bombing Cambodia, and it was time to discuss what was happening to the world. After a few minutes of a rather tense beginning, the convener called a few fellow students to the podium and there was considerable whispering. The leader then looked straight at me in the back where I was sitting, and said they wanted me to run the meeting—they were not used to doing that sort of thing. I was simultaneously surprised and flattered by their confidence, but it was an easy meeting to chair. To a woman and to a man, we all agreed that the war was misguided, and worse, immoral; it had to be put to a stop. We passed a resolution that took the form of a telegram which I was to send to President Nixon immediately, but leading up to this resolution was a remarkably good discussion not only of the war that outraged us all, but other matters concerning student needs and frustrations. It is the only time in the long history of the Biology Department where we had everybody, at all levels, meet in the same room. Catastrophe and the emotion that goes with it brought us together in a way that untroubled times never do. It was a moment we shall all remember for we were grateful to be unified. It is a pity that only something quite dreadful can do this, but then it was a soothing antidote to a horror we were all feeling.

Some of the demands students were making were not realistic. In some departments there was the demand that at the beginning of a course the students meet and decide what they will be taught. I received such a delegation and was told that was what we had to do in biology. I urged them to think about what they were saying for a moment: how could they decide what should be taught before they knew enough biology to make a sensible decision. I said, "Look at a more extreme case: how could students in a atomic physics course provide any kind of a sensible outline to a faculty member teaching the course before the students had learned some atomic physics?" Fortunately, they saw my point. Some time after that the premedical students came to me and said that we should abolish grades; people should be captured by the subject, and not grub for grades.

I told them that in principle I could not agree more, but there were some problems with such a scheme. Since they were about to go home for the Christmas holiday, I wanted each of them to get in touch with the admission deans of their local medical school, and ask them their views. I suppose it was a mean trick on my part, because when they came to see me after the holiday they all reported that all the deans had also agreed with the virtue of the principle, but the student should not bother to apply to their medical school. It would be impossible to evaluate the student, and they had so many good applicants that they would have to chose among those that had grades. Again a commendable idealism gets blown to bits by reality. The good part of all these ideas, even the abortive ones, is that they changed the University very much for the better. Today students continue to be able to have a say in many things that affect them, something that was not possible before the revolution.

During this period our own children went from Rebecca who was in her early twenties to Andrew who was in his beginning teens. In one way or another they were all affected by the revolution. Rebecca got herself arrested in a civil-rights demonstration trying to confront Lyndon Johnson; Jonathan became a flower child and starved himself so he could avoid the draft by being underweight; Jeremy quietly let his hair grow longer; only Andrew was too young for any visible manifestation of the period. Ruth and I did not find it easy to understand, although we slowly became educated. Oddities in dress and behavior seemed the norm for other people's children, but it took on a different meaning when they were one's own.

One day our gentle Jeremy and I were towing a dying car to the garage, and what we did not know was that it was illegal to tow with a rope—it had to be a stiff rod. It was early days of the changes and Jeremy looked exceedingly hippie with a beard and a ponytail. As the policemen who stopped us got out of their car, Jeremy said to me in a soft voice, "You'd better let me handle this, Dad." Stunned, I acquiesced with considerable misgivings, but the policemen just politely told Jeremy not to do it next time. I was the one who did not have it all worked out. Ruth adjusted much faster, but we both made it in the end, and what is much more important, the children survived it without a visible scar.

I can remember a student coming to see me during that tempestuous period and asking me if I would answer a few questions for his sociology senior thesis. I agreed, and he asked me my attitude on all sorts of current issues concerning his generation. I answered as best I could, and he said he could not understand it; my answers were quite different from the previous faculty members he had interviewed—I seemed to be much more broad-minded. I asked him if the others had teen-age children and he confessed he never thought of asking. If one is not learning from students, one is learning from one's own children.

ECOLOGY AND EVOLUTIONARY BIOLOGY
COME INTO BLOOM

The beginnings of ecology and evolutionary biology at Princeton just happened. There was no great vision, no great plan, no great master strategy, yet very suddenly we were one of the leading institutions in those fields. I have often asked myself what led to this extraordinary rapid leap forward. Here are some of the factors involved.

In the first place there was a general feeling within the University that we had to escape from the molecular wars; a requirement for moving forward. Everyone also realized that biology was changing and rushing off in new directions which reached broader boundaries than before. In practical terms this meant a change in how the departments within the University were organized. The first step was simply to separate the newly forming ecological group and from molecular biology, but there were still many from the old biology department who were working on problems of development, cell physiology and neurobiology with increasing involvement in their molecular aspects. This meant that molecular biology and the rest of biology were overlapping. To solve this difficulty—and it was solved in a most ingenious way—for one year there was a mega-department and during that year each individual faculty member could decide where he or she belonged. Even though I always considered myself a developmental biologist, my affinities to problems of evolution were strong, and I decided to join the non-molecular group. It meant I could keep my old office which I loved, and Dan Rubenstein, the much admired new chairman, let me continue there long after I formally retired in 1990. I occupied that splendid room for 64 years.

There was another reason for the blossoming of ecology. There was a sudden, World-wide concern with our environment. With the population continuously growing, and with the finite nature of our resources—food, water, energy, and so forth—matters that concern us all. Not just biologists, but everyone was awakened by these realities. This was a major shift in attention that was a World-wide phenomenon, and what were biologists doing about it?

Ecology was an ancient discipline with a stellar reputation and that was true mid-century. There were many notable contributors, but per-

FIGURE 10. Dan Rubenstein. This photo captures the idealism of a new faculty member who has yet to be asked to be Chair of the Department (photo courtesy of King's College, Cambridge).

haps the most notable was G. Evelyn Hutchinson at Yale, not only for his own stimulating and original ideas, but those of his students that followed him. And a man of tremendous charm and infinite erudition. Ecology suddenly became a key discipline. It blossomed at roughly the same time as the molecular biology although its emergence was less spectacular but just as triumphant and as durable.

There is a big difference between the two. Molecular biology got a send-off like the firing of a cannon with the discovery of the nature of a gene and how it replicated followed by all sorts of satellite discoveries on how it carried out its job of directing the synthesis of specific proteins. It was a great reductionist drama of enormous consequences that keeps unfolding and will continue to do so for many years to come. Ecology had no firing of a cannon, but rather the fusion of a massive array of facts accumulated over the years, followed by the discovery of new ways of organizing those facts to give them new meaning, new insights. Often those new ways were given a big forward boost by the use of mathematical models. The great mass of facts could now be reframed as orderly principles providing bright flashes of insight into what had been a hopelessly complex morass of ecological details. I can remember having a brief encounter with a very emeritus professor of ecology from elsewhere who maintained that the glory of ecology lay in the multitude and complexity of the facts, and that it was heresy to try and simplify and bundle those facts into explanatory models. Such misguided efforts simply removed one from the realities of nature where complexities reflected the true natural world around us. (He also claimed that all genetics was bosh because it was originally done in milk bottles and that was hardly nature. I had just received my undergraduate degree and I came close to instant terminal apoplexy.)

Both the reductionist approach of molecular biology and the holistic approach of ecology and evolution go hand-in-hand in biology; both are essential. It is looking for the big picture, for the generalizations that underlie the major processes of ecology and evolution on the one hand and the molecular details of all life processes on the other. Holism and reductionism are both required ways of thinking about biology; the two essential ways look through different ends of the telescope.

Robert MacArthur was already leading this new path of glory when he came to Princeton from the University of Pennsylvania. I do not remember the details of why he came, but my vague recollection is that my colleague, Colin Pittendrigh, who had become our Dean of the Graduate School, heard that Robert might be movable and immediately, much to everyone's delight, pursued the matter. Soon he was with us, and was soon joined by additional new faculty to make a powerful center of population biology, as it was then called. At the time he was in his thirties and already was considered the leader of the new wave of interest in the ancient subject. What he had done, before he came, was to show that through the use of mathematical models one could get simplifying insights into the great complexity of nature. This was something championed by his thesis adviser at Yale, G. Evelyn Hutchinson, but Robert went much further with it. The community of conventional ecologists at the time were very opposed to his work. As I have just said, any kind of simplification was a

FIGURE 11. Robert MacArthur. This photo was taken circa his move from the University of Pennsylvania to Princeton (photographer not recorded).

travesty, because the magic of nature was its complexity, and any attempt to change this, especially one using incomprehensible mathematics, was a blasphemy. I think Robert rather enjoyed his role as a counter-culture figure. He was the first to say his mathematical models would be replaced by better ones, but even his earliest attempts shed a large amount of new light. The models were simplifications, but as he said, "Where would physics be without frictionless pulleys."

Immediately some new appointments were made, one of which was Henry Horn, an exceptionally imaginative and interesting person whose work on the geometry of trees became a classic. There was also a sudden influx of first-rate graduate students, postdoctoral fellows, and visiting scholars. Almost overnight Princeton became a Mecca for modern ecology and evolutionary studies. I remember with some amusement the tremendous scorn from the biochemists—all ecology was nature study and soft science, not real science. There was consternation when Robert was elected a member of the National Academy of Sciences, and the biochemists decided to take another look. They asked him to give them a seminar to explain what it was all about. Naturally he did not convince anyone, and he received a very bristly set of questions at the end. Instead of getting mad, he seemed to revel in it. Finally someone asked him why he thought any of the things he was working on were worthwhile. He smiled and said that was rather like asking someone why they thought the music Beethoven wrote was worth listening to.

His first office was right down the hall from my laboratory, and his first house was a hundred yards from ours, so inevitably we began to see quite a bit of one another. I liked him from the start, both for his crystal mind and his inner strength. He was very reserved and never said how he felt, and somehow he made one feel one should not say how one felt: those were things that should be understood, not expressed. He was a man of few words, and when he did speak he had a slight impediment, a catch in the beginning of his sentences. If he approved of something one had done or written, he never oozed diplomatic praise—he would let you know in quite indirect ways. If something was wrong he would tell you immediately. We also had quite a bit of communication on family matters, because he and his wife Betsy also had four children.

What was especially important to me was learning about his ideas, which were quite new to me at the time. He asked me to read a manuscript he was writing, and I remember going over an early draft of the famous book he was doing with E. O. Wilson to be called *Island Biogeography*. It made a big impression on me. I think I was even able to help with the book because Robert had a tendency to say things with as few words as possible, no doubt because of his mathematical training, and that did not always make for easy reading. I was a help because I was able to put my finger right on the things I did not understand, all of which were fixed with a sprinkling of added words.

Our unstated friendship grew and we decided we were too cooped up, and since we both liked to walk, we should play hooky once a week and go for a good walk (with binoculars). We settled for Thursday afternoons, and at the stroke of noon, we would rush off in a car, with sandwiches in our pockets. We went to a variety of places: The sand dunes of

Island Beach, trails in the woods, flat miles along the Delaware and Raritan Canal, sandy roads in the Pine Barrens. He was a gifted field biologist and nothing went unnoticed. He could identify a bird from minimal cues, and he was always right when we got a better view.

I am not sure how aware I was of it then, but it is plain to me now that it was a period of expansion of my horizons. I was always interested in big questions, and especially how different parts of biology fit together, but I was ignorant about many key things, especially those where Robert was a master. It suddenly was possible for me to see that ecological communities were not just a hopeless tangle of charming natural history, but a beautiful edifice that linked together so many of the things I had been thinking about. It was during that period that I began to appreciate more clearly how everything was connected: development (my own subject), animal and plant physiology, community ecology, and even animal behavior. I began to see that the unity was because all organisms are life cycles, and that evolution by natural selection is what controls the nature of those cycles. I was entranced by those long discussions on our walks, and was profoundly affected by them. It is not so much that I saw anything totally new, although there were many new pieces: it was more that it all became more organized and into sharper focus. The living world was made of life cycles, each one of which involved genes giving off their signals, their instructions, that were carried out by a whole cascade of living processes: development, the functioning of the organisms, that is, their physiology, and in the case of animals, their behavior, and how they became organized in communities. Those walks were an important part of my life.

At six o'clock one Sunday morning the telephone shot me out of bed. It was from Alan, one of the MacArthurs' sons, who said his dad told him to call, that he was in tremendous pain, and would I come right away. I jumped into my trousers and was there in no time. I knew Betsy was away visiting her mother. Robert had an excruciating pain—it seemed to be in his kidney, and I assumed it was a kidney stone. I called for the ambulance and bundled him off, and reached Betsy by telephone. By the next day Betsy told me it was cancer of the kidney, which had spread everywhere. He was told he had about a year to live, and his bad kidney was removed immediately.

It was a chilling moment, but he pulled us all through. Right after the operation he was completely in charge of himself, would force himself, with his hand on my shoulder, to walk up and down the hospital hall to build up his strength as quickly as possible, and saw to it that no one talked nonsense to him. He told me when we were alone that this sudden change in his life was a huge shock, but that was the way it was—it had to be accepted. His recovery from the operation was quick, and he did not seem different in any way for some time. He decided that he would write a book summarizing his ideas.

As he began to fail our Thursday afternoon walks would become increasingly modest, with frequent stops while he would sit on a log to catch his breath. Eventually we would make them in the car and park in some lovely spot to admire the colors of the turning leaves. It became hard for him even to get into the car, but he loved the sights of fall, and sitting in the car munching sandwiches, our conversation lost none of its spark. He knew, one dreary week in early November, that I was going to take our youngest, Andrew, on a college tour. I went to see him the evening I got back. He wanted to know all about the trip for he was very concerned about the education of his growing children. There was special feeling about that short visit that I did not understand until early the next morning. Betsy called to say that he had died during the night.

He was an intellectual prince at a very early age, and by the time of his premature death in 1972 his impact had been enormous. His view of nature had become the new Establishment—in a very short time he had created a revolution. A few years after his death I remember talking one evening to a young, and exceedingly bright (and successful) British molecular biologist. He asked me if I had known MacArthur, and I explained how we had come to be friends. He said that he was a senior at the University of York when Robert had died, and felt so emotional about it that he almost became an ecologist to carry the torch. It was only after he calmed down that he understood that mathematical ecology was not his natural bent.

Robert was able to finish his book (*Geographical Ecology*) before his death; it was an important work that made a fitting envoi.

The group had now expanded with the addition of Egbert Leigh, Henry Horn, and John Terborgh, but how to replace Robert? This was a

FIGURE 12. Biology Faculty mentioned in the text, late 1980's, caricatured by Henry Horn as their research organisms (except for "Chairman Cox"). *Top Row:* John T. Bonner, Ted Cox, Jim Gould, and Henry Horn; ... *Bottom Row:* Bill Jacobs, Bob May, Dan Rubenstein, and John Terborgh.

matter that really occupied a lot of thought on Robert's part during that last year, and it is an interesting tale. Bob May, a theoretical physicist from the University of Sydney in Australia was on leave at the Institute for Advanced Study in Princeton and he called on Robert one day to express an interest in ecology and was thinking seriously of leaving physics. He also asked if Robert had any problem on which he might like a little mathematical help. This was told to me by Robert, who said he had had similar approaches before and he was immediately wary, but he also was prepared. He pulled a manuscript from his desk which had been totally stuck there for some time for neither he nor others had been able to solve a mathematical difficulty. Bob took it home with him and he had it solved by the next morning, much to Robert's delight, and they published a paper together on it. Somehow this anecdote perfectly reflects Bob's mind and his extraordinary gifts.

Bob was extremely competitive in everything: backgammon, ping-pong, long-distance running—in everything he did. Even in his science, although he was conspicuously generous and helpful to students and colleagues. His generosity and loyalty to students, post-docs, and like-minded colleagues, especially junior colleagues, was particularly admirable. His competitiveness elicited some teasing on the part of fellow faculty and even students, which Bob took with good grace. For instance, when Bob overheard two graduate students talking about running to keep in shape, he challenged them to a mile race—which he won. The race became a semi-annual event, and when graduate students began to win, Bob invited a seminar speaker at the time of the next race—Berndt Heinrich, a gifted scientist and author who happened to be almost a professional runner. Berndt set a new race record, well ahead of the fastest graduate student—who was closely followed by Bob!

This was all later; when he collaborated with Robert he was largely an unknown, except for his mathematical wizardry. They obviously hit it off—it was mutual admiration—and Robert decided to persuade the administration to hire Bob as his replacement. No doubt there are other cases where a professor chose their successor, but they must be exceedingly rare. It is more often the case that making a new appointment is used to clean out the stable. The difference here was that Robert was just into his forties and at the zenith of his faculties and fame, so the administration listened to him and Bob was offered a professorship that he gladly accepted. This had been easy, but then Robert began to worry: was Bob May going to get along get along with the others and was he the best choice? This was a dominant matter on his mind and a frequent topic on our Thursday outings. I kept assuring him that it was a brilliant coup and that Bob already had the respect and liking of his future colleagues, but Robert still worried. He need not have for Bob's intellectual power was different, but quite up to that of his own. And he already had the respect and the friendship of his future colleagues. The result has been that the Department thrived and continues to this day to be a major center for ecology and evolutionary biology. The intellectual leadership of what Robert started is being carried on now by Simon Levin, Steve Pacala, Dan Rubenstein (our chairman), Jeanne Altmann, Peter and Rosemary Grant, Jim Gould, Henry Horn, Andy Dobson, Laura Landweber, Iain Couzin, Lars Hedin, Bryan Grenfell, Andrea Graham, David Wilcove, Peter Andofatto, and a number of younger members who are blasting their way into the future.

When Bob May left us I could not detect a whisper of decline; the world continued to give us high marks. That is not to say things were

FIGURE 13. Ecology and Evolutionary Biology Faculty mentioned in the text, at the turn of the Millennium. *Top Row:* Jeanne Altmann, Andy Dobson, Jim Gould, Peter Grant, and Rosemary Grant; ... *Bottom Row:* Henry Horn, Laura Landweber, Simon Levin, Steve Pacala, and Dan Rubenstein (photos from Ecology and Evolutionary Biology at Princeton University 1999 brochure to attract potential graduate students, plus Rubenstein by Nancy Rubenstein).

identical: Robert and Bob and those that followed, like Simon Levin, are very different people. What was the same is everyone got along and there was much discussion and interaction between the members of the group and a number of joint projects were generated and published. What was different was the way Bob approached any problem. He was fast and very good at instantly getting to the core and doing it in such a way that his analysis led to a clear path to what action should be taken.

The faculty meetings were different, albeit equally successful. Sometimes I was the butt of the fun. In our meetings of the senior faculty to discuss salaries for the junior faculty we often would have to shift back and forth between percent increase and actual dollars. I had a little German pocket abacus of which I was very fond and put it to good effect as I had in previous years. At the first such meeting with Bob he roared with laughter because he could in his head do all the needed calculations at twice the speed, so we switched to the new instrument.

Along the same lines he showed mental agility in another way that has always amazed me. Quite often at a meeting someone would make what to me was a totally incomprehensible point simply because it was so

FIGURE 14. New Ecology and Evolutionary Biology Faculty for the New Millenium (as of 2012). *Top Row:* Peter Andofatto, Iain Couzin, and Andrea Graham; ... *Bottom Row:* Bryan Grenfell, Lars Hedin, and David Wilcove (photos: Andofatto by Ladan Mehranvar, Couzin by Gabriel Miller, Wilcove by Jon Roemer, and the others by Henry Horn).

badly expressed. Bob would immediately intervene with, "what your saying is...." Followed by a short, crystal clear statement from which everybody benefited, especially the person who had made the original garbled point. His verbal facility was quite astounding.

The Princeton University Press—it was really Bob's wife Judy who was one of their editors—decided that the Press would put out a facsimile edition of Darwin's *The Descent of Man* and asked Bob and myself if we would write an introduction. We agreed and I spent much of one summer working on it. A lot of reading was involved and attempts to understand where Darwin's remarkably modern ideas fit in with current ones. When I returned from Cape Breton in the fall I asked Bob about his progress and he confessed he had done nothing on it and Judy was hounding him. But I came to realize in retrospect that he must have done something; he was setting the scene for a spectacular finale. He continued to tell Judy and me not to worry; he would meet the deadline. Then one day just before it he holed up at his house with a recording machine and with the pages I had written and by the end of the day he had finished. All that I had written and even more of his own had been neatly integrated into one coherent essay, making my drawn-out summer efforts seem rather pathetic. He was a magician and everybody was happy.

There are an infinite number of Bob May stories and things went rather pallid after he left. It is not only his distinguished science, but all the activities of his fertile mind. For instance he prided himself in being a master tactician, so he always had advice on what was the expedient way of doing something, which he would shower on the deans of the University and certainly his chairman, which sometimes meant me. In general I found his advice wise and helpful, but inevitably we disagreed on some particular issue. One year when I returned as chairman to take the place of the chairman who was on sabbatical, Bob came to me asking me to do something right away because the real chairman might not want to do it. That returning chairman was a good friend and I felt confident he would agree with what we wanted to do, and I saw no need of doing it behind his back, before his imminent return. I wanted him to be part of a good decision. My answer was a simple "no" (gently delivered) to Bob's request and he became quite furious and before he bounced out of my room his parting shot was to the effect that everyone thinks you are a nice guy but they have not looked underneath. His agile mind was capable of doing wonders in all directions, but occasionally I liked some of my ideas best.

One day he announced that he was going to Britain to take up a professorship at Oxford. In some ways it was no surprise: he had many British friends and I could see he was getting restless. There were new worlds to conquer, and conquer them he did. He continued to be a star in mathematical ecology with many first-rate contributions, but then there were other achievements that clearly gave him deserved pleasure. Let me list some of them: he won the Crawfoord Prize, a signal achievement of Nobel Prize proportions; he was knighted to become Sir Robert; he received the

Order of Merit; he became President of the Royal Society, Her Majesty's Chief Scientific Adviser, and he capped all this and more by becoming Lord Robert May. He has had a truly remarkable career all the way from solving Robert MacArthur's math problem to become a peer of the realm. All the product of a versatile and truly remarkable brain.

THE FUTURE

I know, from having been around for a long time, that I am hopeless when it comes to predicting the future; I am better in knowing about the past and think of it as possibly a feeble guide to the future. The fact is that I never foresaw what direction biology might take, and was constantly amazed and delighted when it did take a new, profitable turn. I have so far in this essay intimated the state of biology, and some of its meandering during the last 65 years. This is my only vantage point to view the future: maybe history does repeat itself and I think there are some lessons from this bit of history. All my predictions for the future during those 65 years were wrong; were blind. So any predictions I might make now would almost certainly be wrong. To begin with, I did not see the coming of Watson and Crick's explosion, nor did I foresee the MacArthur revolution. They just happened with me gasping on the sidelines.

Beyond these major events, what are some of the changes that have occurred in the field of biology since 1947? And where are we headed? One significant bit of progress that I have already discussed is the increasing importance of instruments and techniques and how they made it possible to do the impossible. We went from ancient, cumbersome balances and ordinary microscopes to simpler to far more efficient balances and to microscopes of great complexity and great powers, including electron microscopes that opened up new worlds to be seen for the first time. And it was not just the wonder of the new instruments, but the new techniques that went with them. This may be seen in the nature of some of the recent (and ancient) Nobel Prizes where the reward has not gone to those who discovered a new biological principle, but to those who invented a new technique. Consider the invention of the PCR method that has come into universal use to magnify small traces of DNA, a technique that is not only of prime importance in biology, but all of police work as well. The Nobel Prize was awarded to Kary Mullis for this discovery in the 1980's. In some ways it is a modest, technical discovery, but its impact has been enormous and that is what has made it worthy of the prize. There are other similar examples where new techniques have led to great advances. This certainly is not true of all the prizes; others have been for new, and sometimes revolutionary,

FIGURE 15. John Bonner in his office in 2009. Additions since the original 1948 office described in the text include: water lines, telephone, laptop computer, and notebooks full of new ideas (photo by Denise Applewhite).

principles—great leaps forward. Here I want simply to make the point that technological advances have gone hand-in-hand with advances in principles and ideas in many areas of biology. And no doubt that will continue to be the case as it has for science in general over the years. The invention of the telescope centuries ago and its continuing perfection has allowed us to peek at the universe. And the invention of the microscope by van Leeuwenhoek and others in the eighteenth century has led to many of the wonders of modern biology.

The use of mathematics in biology goes way back. Galileo, for instance, made use of mathematics to illuminate numerous physical principles, and as D'Arcy Thompson showed they are biological principles as well. Mathematics played an important role in the 1930's in population genetics and, as we have seen, has continued to figure prominently in the advances of ecology and evolutionary biology.

I suspect that the effects of technology and the profitable use of mathematics will continue to play important roles in our progress in the years to come, but it is unlikely that they will in themselves open new vistas, new paradigms in the sense of Thomas Kuhn. Those paradigms will spring from bold new ideas, spawned from unusually gifted and imaginative brains, as they have in the past. The use of mathematics in itself is not enough to open new vistas; what is needed is the ability to see the hidden idea and whether mathematics is needed to exploit it is a secondary matter.

I wish I could be here to witness the next 65 years, for I fully expect the Department to continue to bloom.